The Scottish Brick Industry

Moses Jenkins

Castlecary Brickworks of John G Stein and Co. Ltd., 1927.

Text © Moses Jenkins, 2018.
First published in the United Kingdom, 2018, reprinted 2025,
by Stenlake Publishing Ltd.
Telephone: 01290 551122
www.stenlake.co.uk

ISBN 9781840338010

Printed by:
P2D
1 Newlands Road,
Westoning,
MK45 5LD

The publishers regret that they cannot supply copies of any pictures featured in this book.

Acknowledgements

The majority of the photographs in this book were taken by Professor John Hume to whom industrial archaeology in Scotland owes a great debt. Those on pages 5, 10, 11, 12, 13, 15, 17, 18, 19, 23, 30, 31, 32, 36, 37, 38, 39, 41, 43, 44, 45 and 46 are all taken by him and © HES. Reproduced courtesy of J R Hume. A number of other images were taken from the Scottish Industrial Archaeology Survey; those on pages 8, 9, 14, 27, 28, 39, 47 and 48 are all © courtesy of HES (Scottish Industrial Archaeology Survey Collection). The photos on pages 21, 25, 26 and 42 are © HES (Aerofilms Collection). Page 22 is © HES (John Dewar Collection). The photograph on page 48 is © HES (Francis M Chrystal Collection). Those images on pages 6, 7, 8, 24, 33 and 35 are © Crown Copyright, Courtesy of HES. The photograph on the back cover is reproduced courtesy of Gavin Taylor. I am also grateful to all those who took the original photographs and also to Michelle Andersson at the HES photo library for helping get copies of them and ensuring the acknowledgements were correct. Thanks also to the staff at the various archives and libraries where the book was researched, to Dr. Gerard Lynch for many hours of learning related to bricks and to Mark Cranston for running his excellent website. Lastly, this book is dedicated to my beloved wife who makes all things possible.

Further Reading

There are several interesting books for those who want to know more about brickmaking in Scotland. This includes:

Oglethorpe M. and Douglas G., *Brick, Tile and Fireclay Industries in Scotland*
Sanderson K., *The Scottish Refractory Industry*
Quail G., *Garnkirk Fire-Clay*
Bremner D., *The Industries of Scotland*

There is also a wealth of information on Scottish brick history on the website www.scottishbrickhistory.co.uk run by Mark Cranston.

Introduction

In the 19th and 20th centuries, Scotland had an extensive brick making industry which could be found throughout the length and breadth of the country. Over 300 brickworks are known to have been operational in Scotland at one time or another and this is likely to be a considerable underestimation. These ranged from small, rural works which would have supplied only local needs to very large industrialised plants, linked to the rail network and supplying bricks nationally. At the most advanced level Scotland possessed several refractory brickworks which produced material for the lining of furnaces and other industrial plant, products of sufficient quality to see export throughout the world. Despite its extent the brick industry has remained largely under-appreciated. This book seeks to make readers aware of something of the scale, spread and importance of this now almost extinct industry and the leading role which Scotland played in many developments within brick making throughout the 19th century.

Although the use of fired clay bricks was briefly practiced by the Romans in Scotland, following their departure it was not until the 17th century that isolated examples of the use of bricks begins to emerge again. A patent was granted in 1643 for the making of bricks in Scotland but does not seem to have been much used. Firmer evidence can be seen in a contract for repairs to Sanquhar Castle which is suggestive of brick making and also in structures which utilised locally-produced brick dating to the 17th century such as Castle Huntly Ice House. The earliest definite reference to a brickworks being established in Scotland was at Leith in 1709. A brickworks was established at Linktown, Kirkcaldy in 1714 by the architects and entrepreneurs, the Adam family, and the industry then spread throughout Scotland. Other brickworks established in the 18th century include Annan, Cupar, Banff and

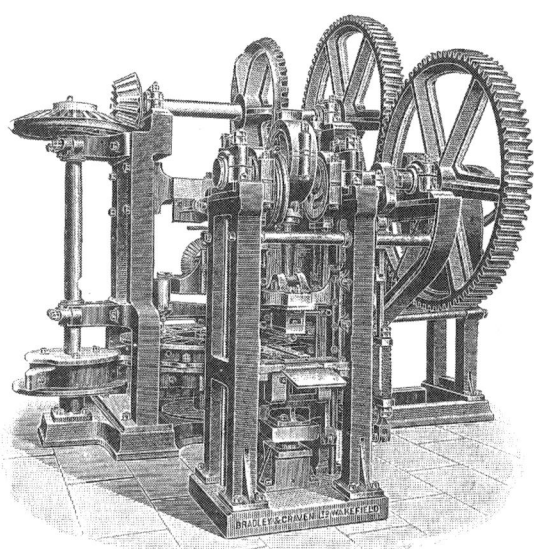

Above: Bradley and Craven Brick Moulding Machine of a type used in several Scottish brickworks in the 19th and 20th centuries.

Duddingstone which was capable of producing 18,000 bricks a week in 1784. At least 32 brickworks were operational in 18th century Scotland although it is likely there were more.

With the advent of the Industrial Revolution both the production and use of brick in Scotland rose dramatically to 15 ¼ million bricks in 1802. By 1840 this had risen to 47 ¾ million. One contemporary observer noted the increase in brick production to be due to both towns and cities expanding and the construction of large engineering works such as railways all of which required bricks in large quantities. Such an increase in production clearly required a brick industry of a significantly larger scale than had existed in the 18th century. Exact figures are difficult to arrive at but at least 300 brickworks were operational in Scotland in the 19th century (although not all at once). During this expansion in the 19th century the geographical focus of brickmaking in Scotland altered significantly. The change from primarily using surface deposits of clay to colliery shale as a raw material had a profound effect on the location of brickworks and the bricks which they produced. Prior to the use of colliery shale Scottish brickmaking was mainly located along the East Coast of Scotland. However, as shale began to be used Lanarkshire, Stirlingshire, Fife and Ayrshire became the focus for new brick making operations.

The 19th century saw many innovations including the first commercially successful brickmaking machine, the Tweeddale Brick and Tile Machine, invented in the mid 19th century. A Scot, Thomas Ainslie, invented one of the first kilns to work on a continuous burning principle, a highly significant advance – he also invented a brickmaking machine. A third significant development was made toward the end of the 19th century by the operator of the Glenboig Refractory Brickworks, James Dunnachie, who developed a highly efficient gas-fired kiln. The 19th century also saw the formation of the most significant specialist branch of brickmaking in Scotland, the manufacturing of refractory bricks, which are manufactured from fireclay and are resistant to high temperatures. They were used in specialist industrial applications such as lining furnaces and gas retorts. The Scottish refractory brick industry was centred around Lanarkshire and Stirlingshire with bricks being exported around the world.

Brick manufacturing in Scotland continued into the 20th century with new brickworks still being established. The link between coal mining and brickmaking meant that when the coal industry was nationalised in 1947 many brickworks were also taken over. These in turn passed to the nationalised Scottish Brick Corporation when it was founded in 1969. As with much of Scotland's heavy industry, brickmaking suffered prolonged decline in the later part of the 20th and early 21st century. There is now only one brickworks operational in Scotland.

Opposite: All brickworks require a source of clay from which bricks are moulded. The types of clay used in Scotland's brickmaking industry have varied widely in their composition but all share certain fundamental characteristics. A good clay for brickmaking is one which contains the minerals silica and alumina together with a flux which will fuse the two together. Various sources of clay were exploited including boulder clay, raised beach clay and alluvium. In some instances clay deposits were interspersed with sand. If too much sand was present, however, the bricks would become brittle when fired. If no flux was naturally present within the clay the silica and alumina would not fuse together and it was termed "foul clay". The photo shows clay being extracted for use at the Tipperty Brick and Tile Works, Aberdeenshire in the 1970s. Clay was largely extracted by hand well into the 20th century, powered excavators being rare. A tramway was used at Tipperty to take the clay from the pit to the moulding shed, a common feature of many brickworks. More substantial examples of tramways for transporting clay existed at Pitfour, Perthshire; Cruden Bay, Aberdeenshire; Sanquhar in Dumfriesshire; the Alloa Brickworks; and the Shore Brickworks in Stirling. Some larger works used stationary steam engines to pull wagons loaded with clay to the brickworks.

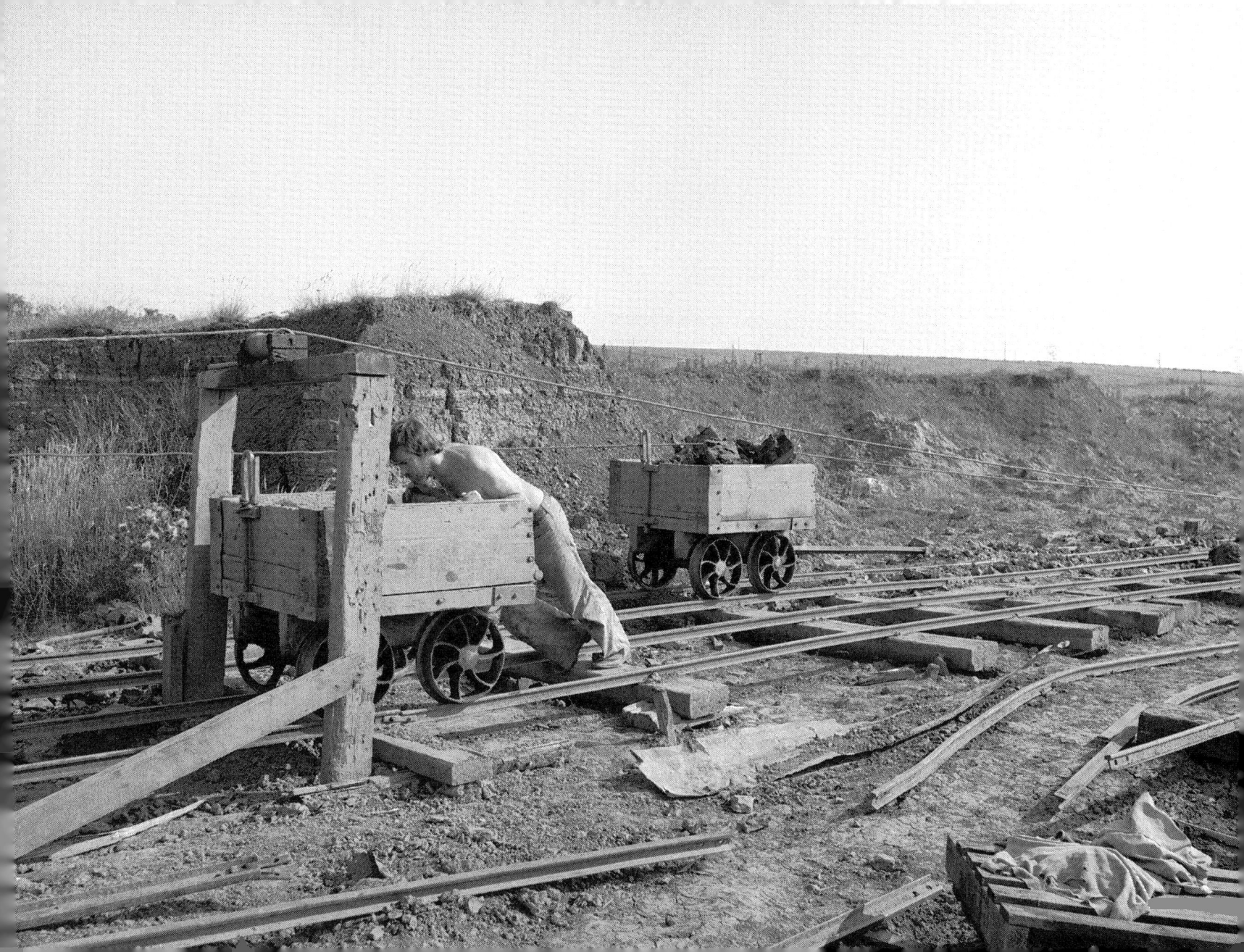

Transportation of clay by tramway, in this case using small hand-powered bogeys at the Blackpots Brickworks, Aberdeenshire, in the mid 20th century. The clay is being dug from a shallow hillside quarry, the material suitable for brickmaking lying just under the topsoil or "overburden" as it was termed. Little scientific analysis of the clay used was practiced until the 20th century, the exception being firebrick manufacturing for which the proportion of silica and alumina was of the utmost importance. An Ayrshire brickmaker in the 19th century described how he would rub a sample of clay between his fingers in order to ascertain its suitability or otherwise. The men who dug the clay were known in Scotland as "holemen" and those transporting the clay as "barrowmen".

In the later part of the 19th century a new material was exploited by the industry. Shale, or "blaes" as it was commonly called in Scotland, was a form of tough clay material which could only be used for brickmaking following advances in clay grinding and milling technology and new moulding techniques such as semi-dry. Blaes was a natural by-product of coal mining and mining companies often established brickworks to make use of it. The Blairadam Coal Company ran a brickworks at Kelty in Fife, the Benhar Coal Company also owned the Niddrie Brickworks and the Faskine Brick and Coal Company had interests in both industries. The photograph above shows a large shale "bing", the spoil heap of shale extracted during coal mining, at Oakbank in Midlothian, the shale in this area being exploited for the production of shale oil, caustic soda and various other products as well as bricks.

Following its extraction the clay was processed. This helped remove stones and other material that would lead to defects in the finished bricks and also helped it gain sufficient malleability to allow it to be manufactured into bricks. This processing sometimes took the form of putting the clay through a pug mill, a machine formed of a large tub or barrel-shaped container with blades internally to chop the clay. The pug mills were powered by horse, water or steam power. Later advances saw the development of pan mills and crushing rollers. *Above right*: A simple horse-driven pug mill. It was used at the Foreland Brick and Tile Works, Islay in the later part of the 19th century. *Above left*: The pan mill at Dunaskin Brickworks, Ayrshire. In this mill clay was ground into small pieces, then crushed by the two rollers and fed through the perforated base of the pan from where a chute dropped it onto a conveyor belt which carried it to the moulding shed where it was pressed into bricks. Pan mills were crucial to the successful use of colliery shale as a raw material in brickmaking as only through the use of more powerful processing machines could shale be made suitable for use.

Opposite: Fireclay, the raw material for refractory bricks, was not extracted in a malleable, plastic state but was a hard material which was mined in a similar way to coal. The "stoop and room" method was frequently employed, the "stoop" being a wide pillar which supported the roof of a mine and the "room" being the area from which the clay was extracted. Holes were drilled in the face of the fireclay deposit being worked and explosives set off only at the end of a day's work to allow dust to settle before the next day's work commenced.

The fireclay mine at Carbrook, Falkirk from the top of the incline at the winding house which brought hutches of fireclay to the surface. This mine opened in 1952 and closed 30 years later, the photograph being taken toward the end of its operation.

During the 18th and much of the 19th centuries bricks were manufactured by hand. This could be carried out using either "sand moulding" or "slop moulding". Slop moulding involved a wooden mould placed onto a moulders table, the mould being wetted first to facilitate the removal of the brick. Clay would be thrown firmly into the mould with the excess being removed by a wire tool known as a harp or a strike. Sand moulding involved a piece of wood called a stock board being fixed to the moulding table onto which the mould box to form the brick was placed. From the late 18th century the stock board could have a raised block attached to its centre which formed the "frog" of the brick, the characteristic indentation found on many bricks, sand being used to dust the mould allowing the moulded brick to come loose hence the name sand moulding. Hand moulding of bricks in Scotland continued well into the 20th century, especially for the production of special shaped bricks and fire bricks. In the 1860s an expert moulder could produce 4,000 to 5,000 bricks a day. The photo shows the hand moulding of a brick at Grahamston Firebrick Works in Falkirk in 1971. In this case rather than sand or slop moulding the mould is being dusted with saw dust, the mould having previously been coated in brick oil. The mould is on the table, a large flat special brick is being produced and there is a pile of clay at the other end of the table. This scene would have changed little from 200 years previously.

Brickmaking by extrusion, Tipperty Brick and Tile Works. Early attempts to mechanise the brick making process generally worked on the principle of extrusion. Clay was forced through a brick – or tile – shaped die producing a continuous brick shaped column of clay which was then cut using wires to the appropriate size. One of the earliest commercially successful brick making machines, The Tweeddale Brick and Tile Machine, was invented in Scotland and was designed to be powered by horse, steam or water power or to be worked by a "man or strong boy". It could produce 15-20 bricks per minute and worked on the principle of extrusion. A foundry at North Berwick produced the Tweeddale which was used throughout Britain. Simple extrusion machines such as this had changed little from the invention of the Tweeddale machine in the 1830s although the method of powering them was by this time diesel engine rather than steam or water power.

A later development was the invention of the stiff plastic process which allowed harder clays and colliery shale to be used as a raw material. The clay or shale was ground to a powder, mixed with water to obtain the required plasticity and then shaped into bricks by machine. A machine manufactured by William Johnston of Leeds for carrying out this process was described as a "Patent semi plastic brick making machine." Two of these machines were bought by the owners of Corsehill Brickworks in Ayrshire and the process became common in Scottish brickmaking from the 1880s onwards. *Left*: A stiff plastic brick making machine manufactured by Bradley and Craven of Wakefield, England, at the Caledonian Brickworks, Larkhall. The conveyor belt in the foreground carried "green" (unfired) bricks to the kiln. This was one of many opened in North Lanarkshire in the 1880s and 1890s to take advantage of colliery shale as a raw material for brickmaking. *Right*: A Whittaker single press and mixer which was used at Jawcraig Brickworks, Stirlingshire in the mid 20th century.

Machines for pressing clay into moulds in what is known as the "plastic press" process came to be used in Scotland later than machines which worked on the principle of extrusion. The first firm evidence for their use was at Allan and Mann's Brickworks in Rutherglen which purchased a brick moulding and pressing machine from Henry Clayton and Co. of London in the 1860s. The machine above is a plastic clay brick press at the Drumpark Brickworks at Bargeddie in Lanarkshire, manufactured by the firm James Mitchell and Co. of Cambuslang who produced a variety of brick making machines. Such machines were not always successful. When the Castlecary Brickworks purchased one it was unable to process the tough local shale, an alternative machine having to be purchased. The works at Drumpark were one of only a small number in Scotland who produced engineering bricks, which were denser, fired at a higher temperature and were thus more durable than standard bricks.

By the 20th century the industry in Scotland was highly mechanised with a number of large, efficient machines in use. Here the extruder and de-airer at Cruden Bay Brick and Tile Works, Aberdeenshire is shown. Machines such as this worked broadly on the same principle as the earlier Tweeddale machine but on a much larger scale. By removing the air from the clay during processing it was ready for immediate use and a much denser and more durable brick could be produced. The Cruden Bay Brickworks was established in 1902 and produced 12,000 bricks a day by 1906. Drainage pipes and tiles were also manufactured at the works, which were extended after the Second World War and by 1960 were producing five million bricks a year. The works closed in 1990.

A number of Scottish firms produced brickmaking machinery in the 19th and 20th centuries, the earliest example probably being Robert Bridges' North Berwick Foundry which, as well as mill machinery and steam engines, manufactured the Tweeddale machine from the 1830s onwards. A later example of indigenous brick making machine manufacture was Blackadder Brothers at the Garrison Works, Falkirk who manufactured "brick and chemical machinery" according to an advert of 1903 and were operational throughout the latter half of the 19th century until the 1930s. Another firm, the millwrights and engineers Thomas Brown of Kirkcaldy, also manufactured brick, tile and pipe machinery in the latter half of the 19th century. A third millwright firm who was involved in the making of brickmaking machinery as well as a considerable range of equipment associated with coal extraction was Hugh Martin between the 1860s and early 20th century. William Brodie, of the Seafield Brickworks near Dunbar, as well as being brick makers in their own right, had an extensive engineering concern manufacturing brick making machinery including pug mils, bruising rollers, pipe, tile and brickmaking machines and also steam engines to power them.

Machines for both clay processing and brick making required a source of motive power. In the 18th and early 19th century this would have taken the form of horse power for the running of pug mills. When the temporary brickfield at Taymouth Castle was extended, a horse-powered pug mill was in use, payment being made for both a "hors mill" and "3 horses for grinding the clay". Water power was also used in some instances although this seems to be comparatively rarer than horse or steam power. In 1856 Tod's Mill brickworks used a wheel propelled by water to power brick making machinery. Other possible locations where water power was employed as motive force include Millands, Auchterarder, Selkirk and Ward's Brickworks, Gartocharn. As technology improved in the 19th century and the cost of engines fell, steam came to be employed for both clay mills and brick making machines. The earliest use of steam power in this industry is thought to have been at the brickworks in Alloa in 1815. The machines which allowed processes such as stiff plastic and semi dry were all driven by steam power in the 19th and 20th century with diesel engines being used later. Above is a stationary steam engine from the mid 19th century of a type which would have been used at brickworks in Scotland into the 20th century.

Before firing bricks it was necessary to reduce their moisture content through drying. This could be achieved in the open air but would be at the mercy of the weather. Where outdoor drying was practised the unfired "green" bricks were stacked in what were termed dykes, with an air gap between each one to allow them to dry. Most brickworks did not dry in the open air, however, and instead had sheds called hack houses, often with louvered sides which could be adjusted to allow more or less air to pass through. Examples of this type of structure could be seen at many Scottish brickworks including Blackpots Brickworks in Aberdeenshire and Inchoonans, Perthshire. Above is the drying shed at Blackpots taken in the mid 20th century. Drying could also be achieved by the use of sheds with heated floors, a particular Scottish term for this arrangement being a "stove".

The next stage in the brickmaking process was to fire them in kilns. Until the mid 19th century the kilns used were fairly small in size and were "intermittent" in their operation. An intermittent kiln is one which is loaded with unfired bricks, the firing is completed and it is then unloaded when it has cooled before the cycle begins again. Intermittent kilns can, themselves, be subdivided into three categories depending on their mode of operation, up-draught, down-draught or horizontal draught. Intermittent up-draught kilns are the earliest and simplest form of brick kilns used in Scotland. The fuel used to fire the bricks in such kilns was either distributed in layers throughout the bricks, or on grates at the bottom of the kiln. Heat travelled up through the bricks being fired with smoke leaving the kiln directly from the top. Down-draught kilns were an advance on up-draught. In these kilns heat and smoke rose to the crown of the kiln which was then drawn down through the bricks being fired, through a perforated floor and out through a chimney. These kilns allowed very close control of firing and a greater uniformity of heat compared to up-draught. Intermittent kilns were either circular or rectangular in shape. Although less efficient than the continuous kilns described on the following page, they allowed close control of temperature and were still used in Scotland well into the 20th century. *Above*: One of the two updraught kilns used at Auchenheath Brick and Tile Works, Lanarkshire which remained in use until the works closed in the 1970s shortly after this image was taken.

In the mid 19th century continuous kilns were invented. They were constructed of a series of chambers. As one chamber was being fired the waste heat and exhaust gases were used to dry the bricks in the next chamber, firing moving from chamber to chamber until all were fired. This was a significant advance and especially useful in the firing of bricks manufactured from colliery shale. Although the Hoffman kiln was the most widespread early continuous kiln, an earlier version, the Ainslie Kiln was invented by a Scot, Thomas Ainslie in the mid 19th century and was used by several brickworks including Heathfield, Garnkirk and Dalquharn. Ainslie later moved his operations to London but it is important to recognise that significant advances in brick making were being made in Scotland in the 19th century. The Hoffman Kiln illustrated was at the Frankfield Brickworks, Earnside, Glasgow, the photograph here being taken in the 1970s. The bricks to the side of the openings were used to brick up the doors prior to firing.

By the 20th century modern brickworks were established which integrated many of the manufacturing processes which were previously carried out in separate stages. Above is the loading bank at the Manuel Works owned by Stein of Bonnybridge in a 1930s photograph. The small bogey with the bricks would have passed through the continuous kiln at the works, following which it would have travelled to the loading bay. On the left the fired bricks are being loaded onto railway wagons. Stein's were a very significant firm in the late 19th and 20th century. Established in 1887 by J.G. Stein when the works at Milnquarter were opened, the company quickly expanded and were at the forefront of the use of new technology. They operated the Castlecary Works, Manuel Fire Brickworks, Anchor Brickworks and Denny and Milnquarter Works and exported firebricks widely throughout Europe and beyond, producing catalogues in several different languages. The works at Manuel were the largest firebrick works in Europe in the latter 20th century producing around 200,000 tons of bricks a year.

Brick manufacture in Scotland began in the late 17th century when a patent was granted to Tobaccos Knowes "for the making of bricks". The 18th century saw the Scottish brick industry develop on a wider scale. One early works can be seen above, the Link Brick and Tile Works in Kirkcaldy, established in part by the renowned Adam family of architects in 1714. The above photograph dates from the 1920s. By the end of the 18th century Scotland had at least 30-40 brickworks, some fairly large such as Duddingstone which was producing 18,000 bricks a week in 1796 and others such as Blairdrummond near Stirling serving only the needs of the local estate on which it was established. By 1779 the brickworks around Portobello and Edinburgh were producing over three million bricks annually, many of which would be destined to form parts of the Edinburgh New Town such as vaulting, internal partitions and as a backing material for stone walls.

Brickmaking was sometimes carried out on the site of a particular major building project. In the 18th century temporary brickworks were established during the construction of Fort George, Inveraray Castle and Gordon Castle. This feature of brick manufacturing continued during the building of large engineering projects in the 19th century, particularly in connection with railway construction. Clay excavated in the digging of tunnels and cuttings could be fed into portable brickmaking machines set up on site. This took place, for example, during the construction of the Kippendavie Tunnel, Dunblane in 1848 during which over 1.5 million bricks were said to have been produced. They were used in the construction not just of the tunnel itself but also stations between Bridge of Allan and Gleneagles. The photo shows Fort George which was an early and significant example of onsite production in Scotland in the 18th century. Two distinct kinds of brick were used in the construction of Fort George, one of a higher quality than the other suggesting considerable knowledge of the properties of clay by the brick makers working on the site.

The earliest firebrick manufactory in Scotland seems to have been that established at Garnkirk, Lanarkshire in 1831. By 1833 the works was producing a wide range of products including ridge tiles, chimney cans and ornamental vases as well as a range of bricks and refractory wares. Even at this early date Garnkirk was clearly positioning itself as a rival to the great firebrick works at Stourbridge in England: an advert in 1833 gave a chemical analysis of both Garnkirk and Stourbridge in order to demonstrate their superiority as well as testimonials from chemists and industrialists including Charles Tennant and Co. of Glasgow. By the mid 19th century, Garnkirk had agents throughout Britain as well as in Baltimore, New York, Philadelphia, Boston and Sydney with bricks also shipped to France, Germany, Russia and the West Indies. The fireclay pits were exhausted in the 1890s and the works closed around this time. Another early firebrick manufactory was established in Falkirk in 1832 which used steam power to grind clay to a fine powder before use in brickmaking. Given its location in Bainsford, it is likely that the output of this works was destined, at least in part, for the Carron Iron Works. In 1867 Glasgow was producing around three million refractory bricks annually, the majority exported to Europe or beyond. *Left*: The Southhook Fireclay Works, Kilmarnock, Ayrshire. The headgear is related to the mine from which the fireclay was extracted. It was originally a coal mine but switched to extraction of fireclay when the coal was exhausted. *Right*: The moulding shop of the Grahamston Firebrick Works, Falkirk. The photograph was taken in 1971 on the last day of production, the works having been operational since 1857.

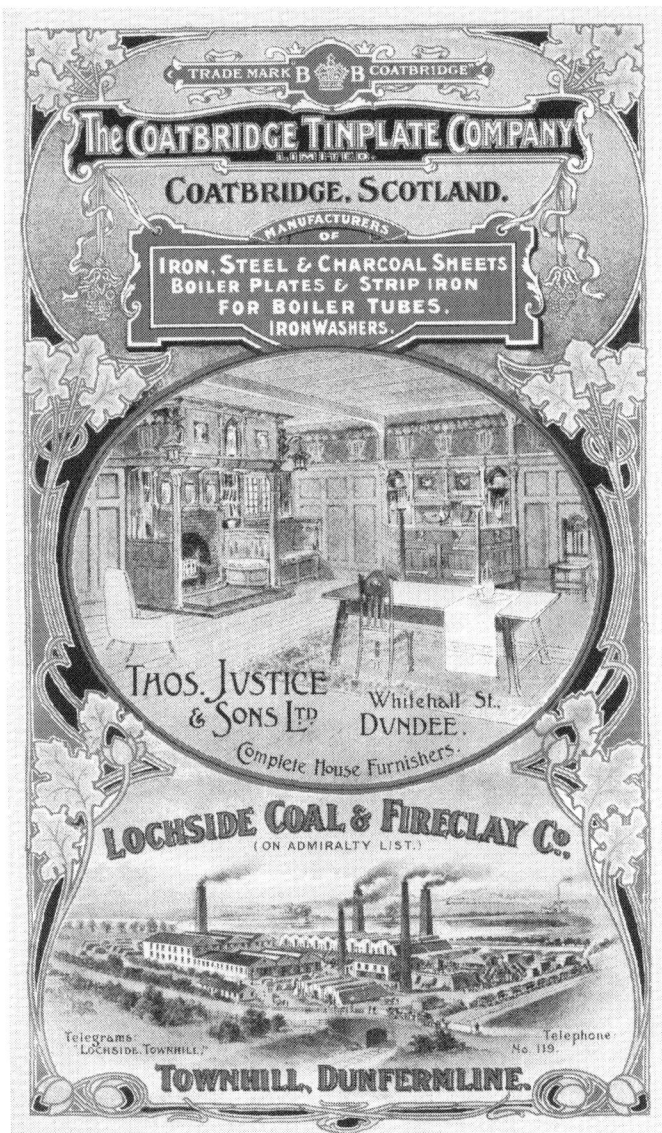

Left: An advert for the Lochside Coal and Fireclay Co. Dunfermline dating to 1905. This was one of several brickworks in Scotland which produced decorative as well as functional items. As well as bricks and tiles, brickmaking firms often produced a wide range of other fired clay items. Chimney cans were just one of the bewildering range produced by the Glenboig Fireclay Company, Lanarkshire in the 19th century. In addition to fire bricks and items for lining furnaces the Glenboig Fireclay Company produced chimney cans, mangers and drinking troughs for cattle. Scottish brickworks could also produce highly decorative products including statuary and urns, and some employed sculptors. The Garnkirk Company had an association with an Italian sculptor living in Scotland, Luigi Isseponi, whose work was highly respected and was exhibited several times at the Edinburgh Royal Scottish Academy in the mid 19th century. They also advertised "Vases, Balustrading and Terra Cotta to any design". The Lillie Hill works in Dunfermline produced highly decorative terracotta sculpture including a Bust of Apollo. The Grangemouth Coal Company's terracotta works produced life-sized statues of Queen Victoria and Prince Albert in the 1850s.

Right: The General Refractories Union Fireclay Works in Glenboig, Lanarkshire in the 1930s. The first refractory brickworks in this area was established in the 1830s at Garnkirk near Gartcosh. The original brickworks became known as the Old Works when one of the partners, James Dunnachie, left to establish the Star Works across the road. The two works were in competition for ten years before merging to form the Glenboig Union Fireclay Company in 1882. Dunnachie made several important advances in brickmaking, including the invention of the Continuous Regenerative Gas Kiln. Bricks from this works were exported to America, Russia, India, the Caribbean, South America and Australia. One curious story tells of an Asian customer exchanging a fine carpet for six Glenboig refractory bricks to be used in the building of a small gold smelter. Various mergers followed in the 20th century with the works closing in 1980 when the firm went into liquidation. It is easy to forget that, as with so much else, Scotland led the world in firebrick manufacture in the latter part of the 19th century with Glenboig being at the heart of this.

In the 19th century brickmaking in Scotland expanded considerably, partly due to advances in technology and partly due to the enormous demand for bricks created by industrialisation and urbanisation. A feature of this expansion was that brick making began to be closely associated with the coal mining industry as coal and shale for brick making were often interspersed with each other. In this 1970s aerial view of Whitehill Colliery, also known as Rosewell Colliery which was owned by the Lothian Coal Company and opened in 1878, the brickworks are in the centre with several small circular kilns and a larger Hoffman Kiln with the moulding workshop and other ancillary buildings adjacent. Workers for the brickworks and colliery lived in the houses on the left.

Despite closure of some long-established works, the Scottish brick industry continued to expand throughout the early 20th century. The Roslin Brickworks (above) was established by the Shotts Iron Company in the 1930s who ran it until 1947 when the National Coal Board took over. The Scottish Brick Corporation became the owners in 1969 and ran the works until its closure in 1977. The photo shows the conveyor and tub way which brought clay into the works. When the coal industry was nationalised in 1947 the National Coal Board came into possession of a number of brickworks. In 1969 they sold off most of these, the majority forming part of the Scottish Brick Corporation. A great many bricks stamped both NCB (National Coal Board) and SBC (Scottish Brick Corporation) can be found throughout central Scotland. As well as the 33 or so works taken over from the NCB the SCB also opened a new flagship works at Bishopbriggs, the Centurion Works, which was said to be able to produce over 72 million facing bricks a year, in 1976. The name Centurion Works was chosen by the workers due to its proximity to the Antonine Wall. The works was truly grand in scale with three continuous kilns each of 28 compartments of 25,000 bricks built to a Dutch design known as Vlamoven. Shale was crushed, ground and processed following the addition of a little water before proceeding to two extrusion machines.

Workers from the Kilchattan Brickworks, Isle of Bute around 1910. Although none are seen here, female workers were involved in brickmaking in Scotland from its earliest beginnings. Their role, along with children, was usually in passing prepared lumps of clay to the expert moulders who were generally male. Brickmaking was often a family concern with women and children aiding male members of the family who worked as moulders. Wages were low and incentives were paid to those who completed a whole brickmaking season. At one Glasgow brickworks in the 19th century "Holemen and barrowmen, engaged in extraction of clay and its transportation were paid 2d per hour, with an end of season bonus of 25s; dykers received 1d per hour with 2s 6d bonus at Glasgow Fair and one eighth of a penny per hour to lie to be paid at the end of the season; machine boys were paid three farthings an hour". Workers at larger brickworks such as the Cleghorn Terracotta Works in Lanarkshire often specialised in particular operations and included chimney top maker (someone who moulded chimney pots), brick burner, brick press worker, brick mixing machine worker, blacksmith, stationary engine keeper as well as several brickmakers all of whom lived in the Terra Cotta Cottages belonging to the company. The sheer length of time which some men worked for the same firm is hard to appreciate today. In 1952 twelve workers at Buick brickmakers in Alloa were presented with silver medals for serving over 53 years at the firm.

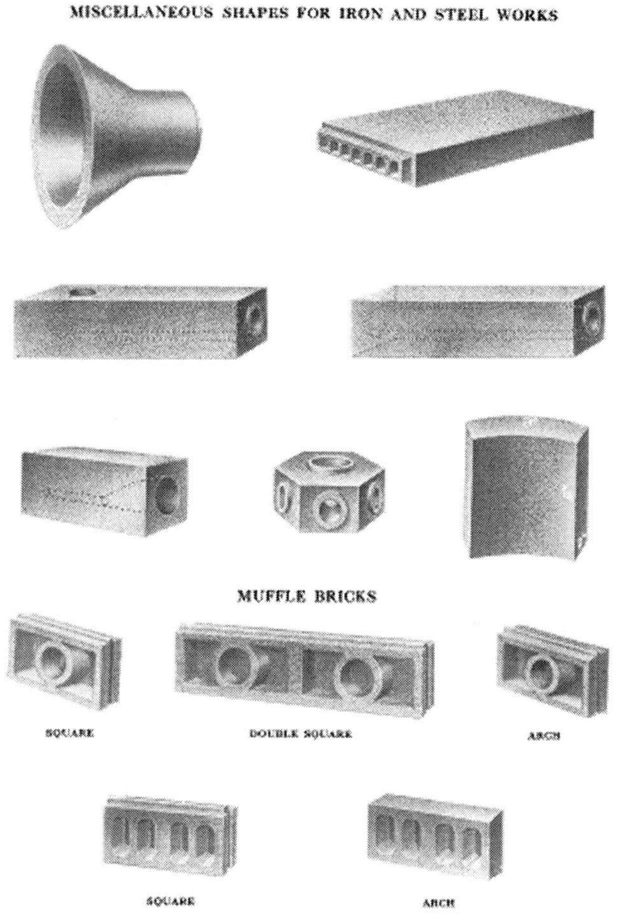

MISCELLANEOUS SHAPES FOR IRON AND STEEL WORKS

JOHN G. STEIN & CO., LIMITED

BURNER BLOCKS

Enlarged Cross Section

Enlarged Cross Section

RECUPERATOR TILES

Patent No. 182073.

MUFFLE BRICKS

SQUARE DOUBLE SQUARE ARCH

SQUARE ARCH

FILLING BRICKS FOR HOT BLAST STOVE

BONNYBRIDGE, SCOTLAND

The refractory brick industry in Scotland made fireclay products of all shapes and sizes for uses in specialised industrial plants. Above are various articles of fireclay produced by Stein of Bonnybridge for use by the iron and steel industry. Firebricks used for the lining of furnaces and gas retorts required to fit together very tightly indeed with a minimal amount of cement between the joints. For this reason they were produced to very accurate shapes, with bricks being produced so as to interlock with each other without the need for cutting. Firebrick manufacturers often also supplied special cement for use in constructing and repairing furnaces and other structures built of firebrick. The above illustrations date to the 1930s and represent just a small part of the firm's output.

The rather sad photograph above shows the partly demolished Buccleuch Brickworks in Sanquhar, Dumfriesshire, with the remains of four circular kilns and the large moulding shed. The works were established in 1889, with clay transported by a tramway, and produced very high quality terracotta facing bricks until closure in 1958 shortly before this image was taken. Dumfriesshire saw brick production established in the 18th century with the Duke of Buccleuch operating three brickworks on his estate in the 1770s which fulfilled several large orders in the years 1774/75, two of which, for 51,000 and 45,000 bricks, were for substantial building projects. Other 18th century brickmaking operations in the south of the country included Tarrasfoot, near Langholm established in the 1790s and Annan in the 1780s. The most significant brickworks in the area was at Sanquhar. It is first recorded as early as 1688 but then closed before going through several operators before commencing the production of high quality facing bricks and paviours in 1889. In the 19th century brickworks in the South of Scotland also included the Paxton Estate Brickworks in Berwickshire, Lamancha Brick and Tile Works in Peeblesshire, Whitribog Brickworks near Kelso and also a brickworks at Cragenholm near New Abbey, Kirkcudbrightshire.

A picturesque view from the 1970s of Sauchie Brickworks with the Ochils in the background. The works opened in the 1880s and was one of several which operated in Clackmannanshire, the most significant one being at Alloa which is likely to have been operational in the 18th century. By 1845 40 people were employed at the works. The works at Alloa were situated close to the harbour and transported their wares along the Forth by boat as well as using the nearby railway. This works was one of the first in Scotland to use steam power for processing clay at the start of the 19th century. Smaller works in Clackmannanshire are known to have operated in the latter half of the 19th century at Coalsnaughton, Dollar, Alva and Blackgrange, all of which are likely to have used colliery shale as their raw material.

The large brickworks at Inchcoonans in Perthshire was in operation from the mid 19th century and made agricultural drainage tiles as well as bricks. It had both round and rectangular kilns which can be seen above. The two round downdraught kilns are to the right along with moulding and drying sheds and stacks of drainage pipes. The works became the Errol Brickworks in the 1980s when this photo was taken but closed in the early 2000s. It was the last manufacturer of handmade bricks in Scotland. Perthshire had a well-developed brick making industry from the 18th century onwards, which continued in operation throughout the 19th century at Grange and Inchoonans in the Errol area, and Redgorton near Stanley. Several new works were established in the 19th century, including Millands and Strathallan works at Aberuthven, Pow Water at Crieff and Scone and Moncrieff both near Perth itself. Moncrieff advertised production of pressed common and circular bricks, likely to be for the construction of farm chimneys in what was a primarily agricultural county. The most significant brickworks in Perthshire was the Pitfour Brick and Tile Works. In the 19th century the Pitfour works was reputed to have installed the most up-to-date plant and soon acquired very extensive business all over the country employing around 50 men and women. It was from here that many of the bricks for the ill-fated first Tay Bridge were supplied although no blame was ascribed to the bricks for the ultimate failure of the structure, which was instead due to deficiencies in the design and the ironwork.

Blackpots Brickworks, near Banff, was established in 1788 and had its own purpose-built harbour. On the left the elevated track brought clay from a shallow quarry on a tramway into the works. The single, rectangular kiln is in the centre. The works ceased operations in the late 1960s around the time the photo was taken. The brick industry in Banffshire expanded considerably in the 19th century, with brickworks established at Tochineal near Cullen and Craigellachie near Aberlour. Other areas of the north east to see 19th century brickmaking included Huntly's Kinnoir Brickworks and Turriff where the Plaiddy Brickworks opened in the mid 19th century.

The atmospheric remains of Foreland Brick and Tile Works on Islay demonstrate how far-flung the industry was. The works were established around 1840 to supply drain tiles and bricks to tenants on the Foreland Estate. The upright masonry piers formed part of the drying shed in which the bricks and tiles dried before being fired. The building to the right was the engine house which housed a horse-powered pug mill for processing clay. The remains of the kiln are to the left. Whilst Islay may seem a remote place to find a brickworks, being far from the industrialised central belt, brickmaking was practised in almost every part of Scotland. The earliest examples in Argyll come from the mid 18th century when a brickworks was established at Limecraigs, Campbeltown as well as a temporary brickfield at Loch Gair, both of which were connected to the re-building of the town and castle of Inveraray. A brickworks was also operational at Slockavullin near Kilmartin in the 19th century. Other West Coast island brickworks were located on Skye at Waternish Bay, at Garabost in Lewis and on Barra and Benbecula. The most significant island brickworks was certainly Kilchattan on Bute. Its production figures in the later decades of the 19th century show over one million bricks were produced in 1896 with the average annual production being between 300,000 and 500,000 bricks.

The brickworks at Brora, Sutherland, seen here in 1889 made use of an outcropping of clay associated with the local coal mine and began operation in 1814, closing in 1826. The works reopened in 1873 and remained operational until final closure in the 1970s. As well as Brora the Highlands saw several other brickworks in operation. The earliest seems to have been the Ussie Brickworks at Maryburgh, Ross-shire, which was operational in the early years of the 19th century and had a water-powered mill for processing clay. The works seems to have ceased operation sometime in the middle of the 19th century but brickmaking in the Highlands was far from over. The Culloden Brickworks opened in 1847 and remained in operation until 1891. A second works around Inverness was established at Millend in the 1880s. There was also one at Allanfearn although its period of operation is unclear and a brickworks at Munlochy near Dingwall was established in 1885. These Highland brickworks, in common with their island counterparts, are all likely to have been fairly small affairs with the exception of Brora.

The earliest brickworks established in Aberdeen was the Seaton Brick and Tile Works. Another significant early brickworks in the north east was established in Peterhead around 1800 making around 250,000 bricks annually. Aberdeenshire also had a well-established rural brick making industry. In the 1850s brickworks had been recently established at Lumbs near Lonmay, Annachie St. Fergus and Downielhills at Peterhead. Other areas which established brickmaking included Blackdog, five miles north of Aberdeen, and Westfield, Auchmacoy. There were also two brickworks established near Ellon by the mid-19th century at Cruden Bay and later in the 1880s at Esslemont. Finally, a brickworks was established at Stonehaven in the latter half of the 19th century. The image above is the Cruden Bay works. A tramway transported clay to the works, which manufactured both bricks and drainage tiles and eventually closed in 1990 making it the last brickworks operational in Aberdeenshire.

Tipperty Brickworks in Aberdeenshire. As with many works in the area drainage tiles and pipes were also produced here as well as bricks, large stacks of which can be seen. The three chimneys are connected to the works kilns.

The most significant concentration of brickmaking in Stirlingshire was around Falkirk, the Forth and Clyde Canal stimulating the industry in a significant way in the early part of the 19th century. Herbertshire, Glen, Callandar and Kerse Brickworks in Falkirk were all located close to the Forth and Clyde Canal. The Kerse Brickworks were located at Lock 3 and the Callandar works at Lock 5. Further along the canal the Bonnyside Brickworks used a tramway to transport bricks to barges and the Horn Brickworks were located at Lock 35 in Drumchapel. Stirling itself had at least two brickworks in the 19th century, The Shore Brick and Tile Works close to the harbour and a second situated near to the railway. Significant expansion took place in the latter 19th century with the exploitation of colliery shale and the establishment in 1886 of the Bleachfield Brick and Tile Works in Falkirk opened for the manufacture of, "Composite and red bricks (machine and handmade) as well as a range of other clay products". The two photos above are of the Avonbridge Brickworks which was established in 1952 but closed in 1977 around the time the photographs were taken. *Left*: Exterior view of the continuous kiln in which the bricks were fired. *Right*: Internal view of the kiln which had 28 chambers in total.

Left: A kiln at Ochiltree Brick and Tile Works being fired using shovels of hot coals. The works were established in 1896 and operated into the 1980s. As well as bricks, tiles and clay pipes were manufactured at the site. A brickworks was established at Kirkwood, Stewarton, Ayrshire in 1779. The industry in Ayrshire expanded in the mid 19th century with new works established at Lanemark in New Cumnock, Springside near Dreghorn, Kames Brick and Tile Works at Muirkirk and Corsehill in Kilwinning. Large integrated industrial sites were created at both Dunaskin and Dalmellington where the same company, Bairds & Dalmellington Ltd., had coal mining, iron working and brickmaking concerns. Ayrshire also saw the establishment of a significant heavy ceramic and glazed brick manufacturing industry in the 19th century. The most significant works of this sort was J and M Craig who operated several Ayrshire works and whose glazed bricks can still be seen throughout Scotland. *Right*: The imposing three chimneys of the Montgomeryfield Brickworks, Dreghorn. This works was established by coal mine owners who utilised shale bings to produce bricks. The chimneys served the four continuous kilns which were operational at the works in the 20th century.

The Montrose Brick and Tile Works shown here was formally known as Puggieston. A tramway carried clay from the clay pits to the works for processing and another carried finished bricks to the nearby railway line for onwards transportation. The works was established in the mid 19th century although it may have had its origins as early as the 1790s. It produced pottery at various times in its existence, some of which survives in Montrose Museum. Brickmaking in Angus centred around Montrose although brickworks also operated at Arbroath and Glamis.

The Niddrie Brickworks supplied many bricks for building in the Edinburgh area. It was owned by the Niddrie and Benhar Coal Company and opened in 1924 to supply bricks for house building in the area. It closed in 1991. The two chimneys are related to the three Hoffman kilns at the works, the image being taken in the 1980s. Brickmaking in the vicinity of Edinburgh has a considerable history, one of the earliest permanent brickworks in Scotland being established at Leith in 1709 with other 18th century works existing at Cousland and Cranston both near the city. Although not on the scale seen in Glasgow, Edinburgh itself had a well-developed industry in the 19th century. In 1863 there were five brickworks within the city: Perceton, Edinburgh Brick and Tile Works, Ferguslie Works, Smeaton Works and Bruce Brickworks. Edinburgh was also served by many brickworks in the Lothians throughout both the 18th and 19th centuries.

West Lothian saw a considerable brick industry in the 18th and 19th centuries. One of the earliest was established at Fauldhouse in the 18th century. Blackness Brickworks opened in the 1830s and exploited a seam of clay 12 feet in depth producing 150,000 bricks annually. Other brickworks established in this period include Musselburgh, East Lothian, and the Smeaton Brickworks in Dalkeith, Midlothian, which in 1830 produced 276,240 common bricks and 19,054 fire bricks. Later, large works were established in West Lothian such as the neighbouring Etna and Atlas Brickworks, Bathville, Armadale, shown in the 1929 photograph. The Etna Works were opened in the 1860s and continued operations until the 1990s. The Atlas works were established in 1882. Both works produced high quality firebricks as well as pipes and a range of other fireclay products. Another large works in the Lothian area was established in 1934 at Pumpherston by Scottish Oils Ltd., which led to the use of the brickmark SOL on its bricks. Waste shale was used as the raw material, over 3,000 per hour being produced by 1950.

Lanarkshire is the county of Scotland which saw the most extensive brickmaking industry in the latter 19th and 20th century. However, prior to mechanisation and the use of colliery shale as a raw material the industry was relatively undeveloped in Lanarkshire. Only the brickworks on the Coats Estate near Coatbridge has been identified in the 18th century although there are likely to have been others. Even in the early 19th century only four further brickworks were in operation, at Carfin, Cambusnethan, Glenboig and Garnkirk, all established prior to 1850, the latter two works specialising in firebrick production. The latter part of the 19th century saw considerable expansion in the brick industry in Lanarkshire and significant concentrations of brickmaking activity existed in Airdrie, Carluke, Motherwell and Coatbridge which all saw at least four brickworks apiece established in the localities. Many of these works were of considerable scale producing many millions of bricks per year. Above is a general view of the Caledonian Brickworks in Larkhall, established at the end of the 19th century. The works' Hoffman kiln is on the right. The small tank on wheels is an oil burner used to light the kiln when firing was taking place.

Brickmaking within the bounds of the city of Glasgow was carried out on a significant scale, beginning in the 18th century with an industry centred around the Calton area where there were deposits of clay. A number of brick makers are recorded in the east of the city with other 18th century brickworks existing at Kingsfield Estate, Clayknowes and Claythorpe, Gallowgate, Glasgow. This had an impact on the housing built in the area, one visitor in the early 19th century noting houses in the east of the city were built of brick, rather than stone, and roofed with pantiles. Above left are the Crown Fireclay Works in the Gallowgate which opened in 1860 and came to specialise in the production of glazed bricks. The moulding shop is next to the chimneys of the two kilns. In common with other areas the production of bricks within Glasgow considerably expanded in the second part of the 19th century so that by 1901 over 40 million bricks were produced in Glasgow every year. Many of these works were fairly short-lived, Gilchrest and Goldie being operational for around 10 years. When clay resources had been exhausted the sites of the works were often built over fairly quickly. In some cases brickmakers also built housing and tenements. The firms of Allan and Mann and A and T Bow had interests in both brickmaking and building. Above right is the Frankfield Brickworks in Springboig which opened in 1899 and had a single Hoffman Kiln.

East Lothian was another area which saw a brickmaking industry established in the 18th century. William Young opened a brickworks at Prestonpans in 1725 with another works at Duddingston opened in 1764 which, by the end of the 18th century was producing 18,000 bricks a week contributing to the reported 3 million bricks produced annually in the Lothians. Other 18th century works in this area included Prestongrange, Musselburgh and Portobello, all of which operated throughout the 18th and 19th century. A number of these works were adapted in the 19th century to take advantage of colliery shale as a raw material. Newly-established brickworks in the latter half of the 19th century included Bankpark in Tranent, Westbank Portobello, and Newbattle near Prestongrange. The Hoffman Kiln in the 1977 photograph of Prestongrange Brickworks is one of the few surviving brick kilns in Scotland. The works at Prestongrange were associated with the local colliery. As well as the large Hoffman Kiln the works also had eleven circular kilns in which glazed pipes were manufactured.

Fife saw the establishment of an early brickworks in 1714, at Linktown, Kirkcaldy, and one at Cupar in 1764. Several works were established in the first half of the 19th century including Methil in 1828 and Inverkeithing in 1831. A works was also established in the 1840s as part of the industrial complex around the Charlestown Lime Works. The first of several brickworks around Dunfermline was also opened in the earlier half of the 19th century at Townhill. Seafield near St Andrews, Dunshelt near Auchtermuchty, Broomlees and Lumphinnans near Cowdenbeath were all established in the 1850s. In common with other industrialised counties a significant number of brickworks were associated with coal mining including Lochgelly, Hill of Beath, Kelty, Eden at Guardbridge and Blairadam which was owned by the Fife Coal Company and used blaes from the adjacent Blairadam Pit as the raw material. Another brickworks connected to coal mining was Wellwood Brickworks near Dunfermline shown on the right. Wellwood was a colliery village and mine in the 19th century with the brickworks being established in connection to this. It lay on the West Fife Mineral Railway which was used to transport the bricks after manufacture.

Many brickworks, particularly in the period prior to the development of the rail network, were situated near a harbour. In the 1790s the Seaton Brickworks in Aberdeenshire used a small sloop on the River Don to bring in coal and carry away brick. In the same period bricks and tiles were shipped from Throsk and the Alloa Brickworks was adjacent to "a convenient Warf". In some instances harbours were specially built for the transportation of brick as at the Invernettie Brickworks near Peterhead. Pitfour Brickworks employed a barge to take bricks along the Tay, a tramway leading from the works to the shore. The Kilchattan Brick and Tile Works on Bute reportedly had "a very extensive trade built up throughout the Western Highlands and Islands and also Northern Ireland" facilitated by transport by sea. This works is shown above in an image dating from around 1910.

The development of the rail network was highly significant to the brick industry and allowed brickworks to transport their wares much more easily. Many brickworks were constructed with transport by rail in mind and a number had dedicated sidings. From their earliest inception Scottish brickmakers used railways to transport the product. In 1832 it was reported that 187 tons of brick had been sent along the Glasgow and Garnkirk line that year and that it was the second highest revenue generator on the line after coal. Dalmellington Brickworks had a branch line allowing bricks to go directly onto the rail network, part of which is shown above with a NCB locomotive at Pennyvenie Mine in the 1970s. Small, rural brickworks also made use of rail to transport their products. Boquahan Brick and Tile Works in Stirlingshire, for example, was linked to the Forth and Clyde Railway by a siding in 1865. Similar examples were the brick and tile works at Bonshaw, Dumfriesshire, the brickworks at Forgandenny, Perthshire, and Cruden Bay Brickworks in Aberdeenshire. A light railway connected the Pitfour Brickworks to the main Perth to Dundee line. A similar light railway ran for 3 ½ miles along the Aberdeenshire coast to connect the Black Dog Brickworks with the Aberdeen to Peterhead line and a tramway connected Tod's Mill Brickworks with the nearby railway line in Linlithgow.